THE SCIE
BOOK OF A~~~~ ~~~~ FACTS

INVENTIONS

Beverley Birch didn't study science at school. She only realised what
she'd missed after growing up in Kenya, graduating from
Cambridge University, and beginning a career as an editor and author,
when she was commissioned to write a children's book on Marie Curie.
Discovering the thrilling facts of Curie's life and work left Beverley with
a particular love of science, and she is now highly-acclaimed for her
inspirational science writing.

The amazing facts in this book were inspired by exhibits
in the SCI~~~~ ~~~~ ~~~~ It is home to many
of the greatest treasures in the history of science,
invention and discovery, and there are also hands-on
galleries where you can try things out for yourself.
If you live in the North of England visit the Science
Museum's outposts, the National Railway Museum in
York and the National Museum of Photography,
Film & Television in Bradford.

Copyright © 1996 Beverley Birch
Illustrations copyright © 1996 Ross Collins

Published by Hodder Children's Books 1996
The right of Beverley Birch to be identified as the Author and
the right of Ross Collins to be identified as the illustrator of the Work
has been asserted by them in accordance with the
Copyright, Designs and Patents Act 1988

10 9 8 7 6 5 4 3 2

A Catalogue record for this book is available
from the British Library

ISBN 0 340 65697 2

Designed by Fiona Webb
Cover illustration by Ainslie MacLeod

Hodder Children's Books
A division of Hodder Headline plc
338 Euston Road
London NW1 3BH

Printed and bound in Great Britain by
Cox & Wyman Ltd, Reading, Berkshire

Sci $\sqrt{m}$

THE SCIENCE MUSEUM BOOK OF AMAZING FACTS

INVENTIONS

BEVERLEY BIRCH

ILLUSTRATED BY ROSS COLLINS

Hodder
Children's
Books

a division of Hodder Headline plc

Contents

Invention and Discovery

All kinds of different people, in different places at different times, have made inventions. Some happened so long ago that we don't know exactly how, when or where. But we do know that behind every invention there is a discovery: people gained new knowledge about the world and realised it could help in the struggle to survive, make their lives easier, save effort.

So it was that someone, once, realised that they could use fire to heat, light and cook. And someone, once, found that heated metal softens and can be shaped and that shaped metal objects can work as tools, pots and weapons. When we discovered how to control electricity, it led to a whole string of inventions which let people communicate faster with each other, drive machines, light their homes . . .

Sometimes the wish to solve a problem comes before the knowledge needed, but people continue to try, undaunted. And there are always some people so full of ideas that they are driven to try them out, just to see, just for the sheer curiosity of it. As you will see in this book, inventions never stop happening. At this very moment someone, somewhere, is looking for new solutions to new — and old — problems.

ANAESTHETIC

Imagine a bloody wooden table. Grim-faced men hold someone down on it. A man leans over the struggling victim. He raises his knife, places it carefully. He ignores the screams, and suddenly, sharply, slices bared flesh ... No, not a horror film, but a real event in hospitals - once. This is a surgeon operating on a patient 150 years ago. Surgeons could try to deaden the pain a little by making a patient very drunk with alcohol. But there was no way then of putting them to sleep, and many died from sheer shock and pain. Speed was the only way to shorten the agony. One surgeon rehearsed by sharpening hundreds of pencils and shaving with his left hand; most got their operating time down to between 54 and 60 seconds.

ONE SMALL SNIFF

Help was on its way, though. From 1842 doctors and dentists in Britain and America searched for an *anaesthetic* - something that could put patients safely to sleep so they would not feel the pain. They tried 'laughing gas' (not strong enough for big operations), and ether (which made people vomit). Then in 1847 a

Scottish doctor, James Simpson, performed an after-dinner experiment. A listener heard the talk in the dining room getting louder and more boisterous, and was shocked by the 'drunken' behaviour. Then a sudden silence - and a thud: Dr Simpson and his two assistants flopping from their chairs to the floor, fast asleep. They had simply sniffed a chemical called chloroform.

'SOOTHING AND QUIETING'

~ Aye, you don't half throw a good party Jim...

Within two weeks Dr Simpson had used chloroform successfully to put 50 patients to sleep. But people were very suspicious. Only after Queen Victoria took chloroform for the birth of her eighth child in 1853 and announced that it was 'soothing, quieting and delightful beyond measure', did it become fashionable. From then on, surgeons could try more complicated operations, and work more slowly and carefully. (*See also* **Antiseptic** *below.*)

ANTISEPTIC

KILLER GERMS

If you were sick 130 years ago, hospitals were about the most dangerous places to be. It wasn't the illness that killed you, but the millions of breeding germs. Wards and operating theatres were filthy. Surgeons wore dusty street clothes during operations, and the stained and gory coats were seen as a sign of how experienced and important they were. They went from patient to patient, ignorant of the hordes of killer germs they carried. Nurses mopped blood with sponges just thrown in a bowl for use on another patient. Over half the patients died because their surgery wounds became *septic* - that is, infected.

AN EXPERIMENT

An English doctor, Joseph Lister, hated seeing so many patients die. He was working in Scotland when he read that Louis Pasteur, the French scientist, had discovered that infections and rotting were caused by germs (a revolutionary new idea then). This gave Lister a brilliant thought. He had seen a chemical called carbolic acid poured on rubbish to kill the rotting smell, and in 1865 he tried it on a patient.

LIVES SAVED

The first time was a terrible disappointment. The man was already too ill, and he died. But Lister would not be put off and tried again, on a young boy's leg-wound. It worked. From then on he used carbolic acid as an *antiseptic* on dressings covering surgical wounds, later he sprayed it on during operations. The death rate plummeted: where 50 in every 100 had been dying, now it was just 3. (*See also* **Anaesthetic** *page 7*, **Penicillin** *page 57*, **Vaccination** *page 115*.)

AUTOMATON

(OR 'ROBOT', BEFORE IT WAS CALLED A ROBOT)

The inventors of the first robot were put on trial as witches. This was in Switzerland in 1733, when Pierre and Henri-Louis Droz first revealed their 'writing machine'. It frightened people because it looked and moved too much like a person, hand-writing messages up to 40 letters long. Fortunately, other people were becoming curious about science and wouldn't accept that the enterprising inventors had done something wicked. In the end the inventors were declared 'not guilty' of any witchcraft.

CONFESS!

Their invention was not called a robot then: this word wasn't used until the 1920s when a playwright used it for the mechanical people in his play. It comes from a Czech word meaning forced labourer (*see* **Robots** *page 89*).

BALANCE

Weighing devices weren't invented to measure useful things like foodstuffs, but for weighing gold dust. The earliest kind of weighing machine was a balance and was invented between 5000 and 4000 BC in Syria and thereabouts. It was a rod, hung horizontally from its central point, with a pan hanging at each end. Polished-stone weights, often shaped like ducks and lions, were put in one pan, and the gold dust in the other, so that they balanced. At that time other goods were just bartered (that means people exchanged one thing for another without using money). About 1500 years passed before people began to weigh food and other goods for buying and selling.

BRAILLE

A thirteen-year-old blind boy transformed the lives of millions of people throughout the world. Over 170 years ago, an army captain showed the young Louis Braille a system for passing written messages between soldiers at night. Each alphabet letter was represented by a group of dots, raised above the surface of the page. You didn't have to see them, you could *feel* the bumps with your fingers.

FASCINATING DOTS

Louis thought there were too many dots to remember, and that writing a word would take too long. But the books at his blind school in Paris were no good, either. They had ordinary letters, made huge so that you could trace the shape with your finger. It was so slow that you forgot the first letter by the time you got to the end of the word.

Dots caught his imagination. So in 1823, just thirteen years old, he set to work.

INTERNATIONAL ALPHABET

For two whole years he struggled with the dots, in every free moment, even in bed in the dormitory at night while everyone else slept. He grouped the dots this way and that, using fewer and fewer and fewer. By the time he was fifteen he had perfected the international alphabet that is still used all over the world by the 42 million people who are blind. It has never been bettered - because it is so beautifully simple. It uses just six raised dots to make every letter, number, punctuation mark and mathematical sign. We call it *braille* after him.

BREAKFAST CEREAL

You can thank Americans with 'delicate stomachs' for ready-to-eat breakfast cereals. The first was in Denver, Colorado just over 100 years ago in 1893. Henry D. Perky met someone who ate boiled wheat soaked in milk every morning to ease his indigestion, and Henry D. Perky liked the idea so much that he produced Shredded Wheat. Around the same time, William Kellogg was looking for ways to make easily-digested wheat-foods to replace bread for the patients at his brother's sanatorium in Battle Creek, Michigan. He left some boiled wheat to stand, and then had the idea of slicing it thinly and baking it. Christened Granose Flakes, it led about four years later to a version made from maize ... Corn Flakes.

BRICK

Brick-making is 12,000 years old, and started a bit like making mud-pies. Around 10,000 BC along the river-banks in the Middle East, people started taking mud from the river and drying the cakes in the hot sun. Mixing straw with the mud made them tougher. Baking the bricks in

an oven to harden them (called 'firing'), came a lot later, around the fourth century BC in Mesopotamia, the ancient name for the land between the Tigris and Euphrates Rivers near the Persian Gulf. Brick houses were stronger, so could be built larger than clay or wooden houses. But today more houses are built with unfired mud bricks than any other material - even though they can be washed away by rain.

CALCULATOR

ABACUS

In the hands of a real expert, a 5000-year-old instrument can be faster than a modern pocket calculator. Ancient Babylonians developed the abacus in about 3000 BC to add, subtract, multiply and divide numbers. You sometimes see simple versions of it in school: it has beads moved along wires to represent numbers. One row of beads stands for the millions, the next for hundred thousands, ten thousands, thousands, hundreds, tens, ones, tenths - and so on. Ancient Hindus, Greeks and Romans, and Europeans in the Middle Ages, all used the abacus, and it is still used successfully in India, China, Japan and Russia.

Beat you again
~ slow boy.

PASCALINE

If you lived in the 1640s you might use a Pascaline to help with arithmetic. It was the true ancestor of our modern calculator. Blaise Pascal, aged only nineteen, invented it in France in 1642, while helping his father collect taxes. He fed numbers into his machine by operating wheels, which turned gears, which moved dials to show the answers in a set of windows, and it could add and subtract eight columns of numbers at a time. Just over 50 years later, the German Gottfried Wilhelm von Leibnitz invented a machine that could also multiply and divide. Right up until the 1970s calculating machines

still worked on these principles of wheels and gears. Then came the electronic versions we use now (*see* **Electronics and all that** *page 74*).

CASH REGISTER

Fed up with customers arguing about bills, an American saloon keeper invented the cash register one November day in 1879. While on a boat-trip to Europe, James J Ritty spotted a machine that counted how many times the propeller turned. It gave him the idea for a machine that could add up the number of drinks sold *and* keep the cash safe. And with his eyes very firmly on the money, Ritty set about making his fortune by selling it to other saloon keepers.

CINEMA

A MURDER MYSTERY?

The first moving films may have been the motive for a murder. The French creator (on his way to show his invention), disappeared from a train, never to be seen again. The films were made in about 1887 by Louis Aimé Augustin Le Prince. In 1890 he was going to Paris to demonstrate his invention. He boarded a train on 16 September, but never arrived. Neither he nor his equipment were ever seen again. The mystery has never been solved to this day.

CINEMA IS BORN

The first film shown on a screen to an audience lasted just eight minutes. It showed workers leaving their factory for the lunch hour. This first showing was just over 100 years ago, on 22 March 1895 - and so the cinema was born. The idea of projecting film onto a screen so that many people could see it was the brain-child of two brothers in France, Louis and Auguste Lumière. Until then, moving films could only be looked at by one person at a time, because you had to peep into a machine called a

Kinetoscope. Soon after the first showing in France, nearly every large country in Europe opened cinemas.

THRILLS, SPILLS AND EXCITEMENT

In 1896 first British audiences were stunned by the sheer excitement of *Arrival of a train at a station* - the train seemed to roar straight at them out of the screen. And if that wasn't enough there was *The Baby and the Goldfish* and *The Family Teatable* to follow! (*See also* **Photography** *page 63*).

I've seen it before I know what happens

COMPUTER

AS BIG AS A HALL

Early computers were vast, heavy, hot, and couldn't remember much. One of the ancestors of modern computers which was developed between 1942 and 1946 in America, was 30 metres long and weighed 30 tonnes. It filled the walls and cupboards of a very large room, and had 18,000 electronic glass tubes for controlling the flow of electricity. These always overheated and needed constantly-moving air to cool them. The computer could only remember 20 numbers and 10 letters at a time; it was difficult to use, often broke down, complicated to repair and used a vast amount of electricity. And if you wanted to set up a new program, you had to re-switch and re-plug for days. But, with other computers being developed at the same time in Britain, it was the beginning ... (*see* **Electronics and all that** *page 74.*)

THE ANALYTICAL ENGINE

The first design for a computer was by Charles Babbage in England over 100 years earlier, around 1834. He called it an 'analytical engine',

and although it was never built, many of the ideas in it were rediscovered by the scientists who did develop the first working computers in the 1940s in Britain and America. Now we realise how clever Babbage's design actually was.

INCURABLE INVENTOR

Babbage was an inventor from childhood. He designed and made: footwear for walking on water, a system for shooting messages in cylinders along cables through his house, an instrument for inspecting the inner eye, a special pen for drawing broken lines on maps, a camper for his holidays - a horse-drawn carriage with beds, cooking and toilet facilities. He also designed, but never made: diving bells, submarines driven by compressed air, an instrument for measuring height above sea level and an arcade game of noughts and crosses. (*See also* **Child inventors** *page 123*.)

Quiz

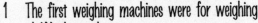

1 The first weighing machines were for weighing
a) Weightwatchers
b) Gold dust
c) Eggs

2 The first bricks were made
a) 100 years ago
b) 2000 years ago
c) 12,000 years ago

3 A Pascaline was
a) A board game
b) A calculator
c) A system of writing for blind people

4 Breakfast cereals were invented for
a) Busy people
b) Sick people
c) Babies

5 The analytical engine was
a) A car
b) A computer
c) A steam engine

6 The first films showed
a) A baby eating a goldfish
b) Workers leaving their factory
c) A murder on a train

7   In 1842 doctors tried to reduce patients' pain with
    a) Crying gas
    b) Laughing gas
    c) North Sea gas

8   The Droz brothers were tried as witches for
    a) Inventing a flying broom
    b) Stealing teeth from graves
    c) Making a writing machine

9   Braille was invented by
    a) A blind army captain
    b) A young boy
    c) Louis Pasteur

10 The first antiseptic was
    a) Chloroform
    b) Sour milk
    c) Carbolic acid

11 Charles Babbage designed
    a) Shoes for walking on water
    b) The first sewing machine
    c) The first computer

12 The cash register was invented by
    a) A French sweet-maker
    b) An American saloon keeper
    c) A Chinese printer

Turning Points and Giant Leaps

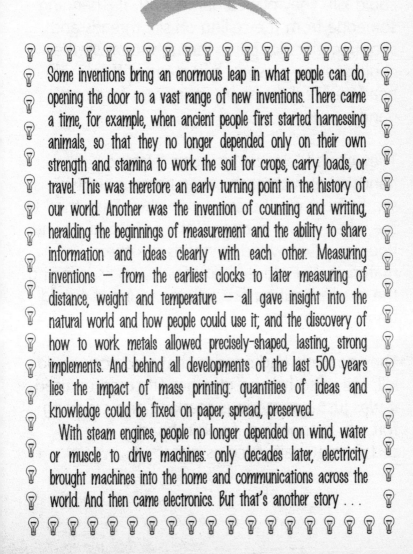

Some inventions bring an enormous leap in what people can do, opening the door to a vast range of new inventions. There came a time, for example, when ancient people first started harnessing animals, so that they no longer depended only on their own strength and stamina to work the soil for crops, carry loads, or travel. This was therefore an early turning point in the history of our world. Another was the invention of counting and writing, heralding the beginnings of measurement and the ability to share information and ideas clearly with each other. Measuring inventions — from the earliest clocks to later measuring of distance, weight and temperature — all gave insight into the natural world and how people could use it; and the discovery of how to work metals allowed precisely-shaped, lasting, strong implements. And behind all developments of the last 500 years lies the impact of mass printing: quantities of ideas and knowledge could be fixed on paper, spread, preserved.

With steam engines, people no longer depended on wind, water or muscle to drive machines: only decades later, electricity brought machines into the home and communications across the world. And then came electronics. But that's another story . . .

ELECTRICITY

SPARKING GAMES

In the 1740s people didn't know that electricity could kill. They played games with it - hanging someone from the ceiling on silk threads and shooting electric sparks through them so they could see the person's hair stand stiff from their head and feel the spark snap from face and hands.

They played 'electric kiss' and 'electric handshake' with each other, passing electric sparks from one person to another. They tried to see what would happen if they all stood on the floor, or all on wooden stands or other materials. They loved exotic party tricks, sending the spark shooting like a flame round the gold frame of a mirror.

They not only had no idea that electricity could kill, they also had no idea how it could be used. It was just a mysterious sparking and crackling force that appeared if you rubbed certain materials, such as glass.

The Ancient Greeks had been the first to notice this strange force - in a natural substance called amber. When they rubbed it with fur, it attracted light objects to it. Their word for amber was *elektron*, and this is where our word 'electricity' comes from.

But by the 1740s, electricity was still peculiar, interesting and fun to play with, little more.

ALL JUMP TOGETHER

The King of France organised his whole brigade of guards to stand in a line, shot an electric shock through them and was delighted to see them all jump in the air together. Another scientist set a three-kilometre-long line of monks leaping. Amazingly, no one seems to have been injured in the experiments.

THE MYSTERIOUS SPARK

A few people were trying to learn more about the spark. They found it was sometimes a snapping brush of blue flame, sometimes short and fierce. They passed it through cannon balls, gun barrels, knitting needles, fire tongs, tea kettles, wood bricks, water, chalk, salt, even the gilt paint on a mirror and the gold-painted flowers on a china cup. They melted brass pins and punched holes in thick board. They also learned how to lead it along a wire into a glass bottle, collect it and lead it out again.

KITE EXPERIMENT

The American Benjamin Franklin was one of the most important of these early experimenters, and famous for flying a kite in a raging storm in 1752 to 'catch' lightning sparks from the clouds. He wanted to know if they were the same as the electric sparks he could make at home. He should have been killed. Not long after, someone else was, while testing Franklin's ideas.

LIGHTNING CONDUCTOR

The kite experiment gave Franklin the idea for the first serious electrical invention - the lightning conductor. Towers, church spires and ships' masts were often burned down in fierce lightning strikes. Franklin's invention was a metal point fixed on top of buildings or ships' masts to draw lightning and lead it safely down the outside of building or ship into the ground or water. It inspired one Frenchman to invent a portable lightning conductor for his umbrella.

CONDUCTING
UMBRELLA
MARK
ONE

ANIMAL ELECTRICITY

It all began with a dead frog. In 1786 Luigi
Galvani, a scientist at the University of Bologna
in Italy, noticed that a dead frog's leg twitched if
he touched it *at the same time* with two wires -
one copper, one zinc. He thought the leg was
full of some kind of special fluid which he called
'animal electricity'.

THE FIRST BATTERY

But some years later (in 1800), another Italian, Alessandro Volta, realised what this really meant, that two *different* metals separated by *any* moist substance would produce a continuous flow of electricity - *an electric current*. At first he used wet paper stacked vertically between metal plates. Then he made a series of cells containing liquid chemicals linked by different metals: the more linked cells, the more powerful the battery.

His invention caused a sensation: for the first time scientists could produce more than just short, fierce sparks that were gone in a flash. So began the long series of experiments that led to the first important electrical inventions (*see* **Turning points and giant leaps** *page 25*, **Electric lamp** *page 32*, **Telegraph** *page 101*).

ELECTRIC LAMP

'In beauty the light surpasses all others, has no smell, emits no smoke, is incapable of explosion, and not requiring air for combustion, can be kept in sealed jars. It ignites without the aid of a taper ... ' This is a journalist's description of the first-ever electric lamp. It seemed almost magical to people then, because all the kinds of lighting then used - gas and oil lamps - had to be lit with a flame.

The electric lamp was created by a Scottish scientist, James Bowman Lindsay, in October 1835, more than 40 years before light bulbs for sale to the public were developed by other inventors (*see* **Light bulb** *page 42*). But Lindsay was just an imaginative man, interested in exploring everything, not in selling inventions to make money. He made his lamp, satisfied his curiosity, and simply moved on to new explorations.

FALSE TEETH

In the 1700s dead men's teeth, taken from skulls in graveyards and battlefields, were used as false teeth. Dentists also used hippopotamus and walrus ivory, and sometimes bone. But they all rotted, so the teeth smelt horrible. Only very skilled dentists could make comfortable false teeth, and only the rich could afford them. Even so, they only wore them for looks, removing them at meal times. Food for the rich and toothless was 'chewed' first with special pliers.

We have to thank a Frenchman, Nicholas Dubois de Chemant, for introducing rot-free, specially-made porcelain dentures to England during the 1790s. He got the idea from an apothecary (chemist) named Alexis Duchateau he had worked with some twenty years earlier.

GAMES AND TOYS

Question: what do the inventors of these games have in common: Meccano (1900), Lego (1955), plastic hula hoops (1958), video games like Space Invaders and Pacman (1970s and 1980s), Dungeons and Dragons (1973) and Dragon Quest (1988), Rubik cube (1979), Trivial Pursuit (1981) and Pictionary (1986)?

Answer: they were young, couldn't find anyone far-sighted enough to be interested in producing their inventions, and had to produce and sell the games themselves. Then they were able to thumb their noses at all the doubters: all these inventors became millionaires on their earnings!

GAME
DESIGN
PROGRAM
☆

LOADING

GUNPOWDER AND EXPLOSIVES

Celebrations always went off with a bang in Ancient China - literally. Sticks of bamboo were thrown into a fire as a useful signal for calling the people of the villages together. The bamboo exploded because the air inside it was heated suddenly. This was the first known explosive, dating from the second century BC. Later (but still over eleven centuries ago), and still perfecting their festivities, the Chinese invented gunpowder for fireworks. There is no record of gunpowder being used to blow up people in warfare until around 400 years later.

At about the same time, knowledge of gunpowder reached Europe. It was the only explosive used until 1788. The first modern explosive, nitroglycerine, came from research on textiles by the Italian Ascanio Sobrero in 1846. The textile research went nowhere, but dynamite (made with nitroglycerine) began to be used for blasting during building work by the Swede Alfred Nobel. For a long time making explosives was very dangerous - nitroglycerine had a habit of blowing up at the slightest jolt, causing the most terrible accidents.

INK

Take beetles (well squashed), soot or lampblack (the sooty deposit left in oil lamps), berries and plant juice. Mix thoroughly with gum and dry into a hard stick. Mix with water before use. This is a typical recipe for ink from the Ancient Chinese and Egyptians. Writing with this kind of ink began about 4500 years ago. Before that, people used a stick to make marks in wet clay.

The Chinese used brushes to write in ink and were the first to print with ink, using wooden blocks. When machine printing began in Europe in the 1400s, ink was still made with lampblack, but mixed with varnish or boiled linseed oil. But it took a long time to dry. Quick-drying inks were developed in the 1700s. Today different printing inks made specially by chemical processes are used to print cloth, plastic, paper and other materials. (*See also* **Printing** *page 69*, **Pens** *page 60*.)

IRONS

Some people nearly electrocuted themselves with the first electric irons. They first appeared in the 1880s, and people needed a warning not to put them in water. Until then, irons were always heated in a fire or on a stove. Not surprisingly, they got rather sooty, and people were used to giving them a good watery scrub. And have you ever wondered where the saying 'having many irons in the fire' comes from? The servants in rich houses would literally have many irons ready heating so that they could get through piles of laundry without stopping.

'Box irons' had a compartment for a heated metal block - a vast improvement, because the surface of the iron didn't get dirty. In parts of the world where electricity isn't readily available, similar charcoal-filled irons are still used. At the beginning of this century electricity was too expensive for most people, so gas irons were popular. However, the user got very hot and bothered because the gas came through a tube and a naked flame heated the base plate while it was being used.

LASER

Surgeons can operate inside an eyeball without ever cutting the eye open - with a laser. This is any device that sends out laser light - a very pure, closely-focused light. Solids, liquids and gases can be made to 'lase' (produce laser light) continuously or in 'pulses' (short bursts). The most common is made by a crystal such as a ruby, stimulated by flashes of very bright light. The first laser, in 1960, was a ruby medical laser, invented by the American scientist Theodore Maiman. Other lasers are produced by gases when electricity is passed through them.

Lasers seal the body's blood vessels as they cut. Doctors can use them to destroy tumours, treat sports injuries, and even remove tattoos (the laser vaporises the dye). Miniature lasers 'read' CDs and videodiscs. And because they can be controlled very accurately, lasers can cut, drill and weld metals and many other substances. Moving at fifteen metres a second, for example, a laser will cut through many thicknesses of textiles, leaving heat-sealed edges that won't fray.

LIFT

FLYING CHAIR

King Louis XV of France felt lazy about climbing from his royal apartments on the first floor of the Palace of Versailles to his mistress, Madame de Chateauroux, on the second floor. So he had a 'flying chair' built in 1743, and had servants haul him up and down on a rope. Hoists to raise loads were invented in ancient times, but people didn't usually ride in them because buildings weren't very tall and, anyway, they feared the rope might break and send them crashing to the ground.

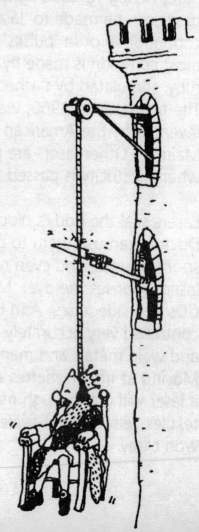

THE SAFETY LIFT: CAUGHT BY THE TEETH

In 1851 an American engineer named Elisha Otis (working in a bed factory and devising ways of moving beds from floor to floor), thought of putting 'teeth' on the walls of the lift shaft. If the rope broke, a bar on the side of the lift would spring out and catch the teeth to stop the lift plunging downwards.

FIRST PASSENGER LIFT

Otis persuaded spectators at the New York exhibition in 1853 to cut the rope with an axe while he stood trustingly in the lift. Luckily for Otis, his invention worked. The first public passenger lift (complete with safety mechanism) was installed in a New York department store in 1857. After that, architects began designing buildings taller than ten storeys high. The skyscraper was soon to follow.

LIGHT BULB

FLAME-FREE

The first electric light bulbs needed a health warning.

> *'This room is equipped with EDISON ELECTRIC LIGHT. Do not attempt to light with match. Simply turn key on wall by the door. The use of electricity for lighting is in no way harmful to health, nor does it affect the soundness of sleep.'*

Thomas Edison developed the light bulb in America during 1878 and 1879; at the same time in Britain, Joseph Swan was developing his own bulb, although they went on sale a year later. (*See also* **Electric lamp** *page 32 and* **Recording** *page 83.*)

HAIRY BEARDS AND FISHING LINES

It took Edison fourteen months and thousands of experiments to get the right *filament* (the loop that glows inside the glass bulb when electricity passes through it). The substance had to burn for a long time without breaking. He tried metal, cotton, flax, Chinese and Italian raw silks, hard woods, horsehair, rubber, cork, assorted grass fibres, linen twine, tar paper, wrapping paper, cardboard, tissue paper, parchment, bamboo fibres, coconut hairs, fishing line, and even hairs tweaked from the beards of his assistants. The best was 'carbonised' paper - paper baked in an oven until only its burnt (carbon) framework was left.

SLOW SWITCH-OVER

But the change from gas to electric light in homes was not as fast as you might think. As little as 70 years ago (1920s), 40 years after light bulbs were first on sale, only about 12 in every 100 homes in Britain had electric lighting. After another 17 years, little more than half the homes had it. Only by about 35 years ago (in 1961) had the number in Britain reached as many as 96 in every 100 homes.

MAGIC LANTERN

Ghosts and demons cavorting across a screen were a popular entertainment by the time the cinema began. They were hand-painted pictures projected onto a screen using a magic lantern. Entertainment of this kind goes back as far as the second century BC, performed to keep Chinese emperors amused. In Europe, these kinds of displays were appearing in wealthy households in the seventeenth century, and were common public entertainments by the eighteenth century. The earliest versions, before electricity, were cut-out paper figurines pasted on a merry-go-round and paraded in front of a candle in a dark room. When moving cinema films began, they were often shown in the same programme as magic lantern shows (*see* **Cinema** *page 19*, **Photography** *page 63*).

Teamwork

In the past inventors have often been individuals who developed an invention to serve a particular need — or because they just liked inventing. Working alone, knowing little — or none — of what is happening elsewhere, they invent a calculating machine, an alphabet for the blind, a game or sport, a battery, ink, a microscope, a sewing machine.

Nowadays most invention is by teams — people who come together to pool their knowledge and skills to solve a problem. Penicillin was developed like this, and teams worked on radar, computers, nylon . . . to name just a few. Marie Curie, Pasteur, Edison, Marconi all formed their own teams to help by bringing new ideas and knowledge and different skills.

But in a sense every inventor, throughout the ages, is always part of a team — because every inventor stands on the shoulders of all who have gone before. Our knowledge of the world has been gathered over thousands of years, pieced together by different people, working in different ways and at different times. But each one adds to our growing understanding, each makes their own contribution to the whole.

Until the day comes when someone, somewhere, puts an important new piece of information into the jigsaw, which gives some inventive and imaginative mind a bright new idea . . .

METAL DETECTOR

The metal detector was invented in a frantic dash to save the life of the President of the United States. In July 1881 an assassin shot President James Garfield, and doctors couldn't find the bullet. Surgery wasn't yet advanced enough to cut him open safely and look for it, so Alexander Graham Bell (already famous for inventing the telephone) took up the challenge. Bell made an electrical device that would make a noise when near metal. (*See also* **Telephone** *page 105*).

As the president's condition worsened, Bell had a desperate struggle to get his invention right. Among other problems, the detector was thrown off-course by metal bedsprings under the president's plush mattress. Before Bell could succeed, the president died. Not from the bullet - from infection. Doctors who were scornful of Joseph Lister's teachings about keeping wounds clean, kept poking unwashed fingers into poor Garfield, looking for the bullet. After his death, they found it, too deep for Bell's machine to detect. The bullet was lodged where it probably would have done no harm - if only the wound had been kept clean! (*See* **Antiseptic** *page 9*.)

if only it had been a canonball...

MICROSCOPE

Amazing creatures, never ever seen before, were discovered with early microscopes. We don't know exactly when the microscope was invented, but it was probably towards the end of the sixteenth century in Holland, and probably by spectacle makers. In the hands of Anton van Leeuwenhoek, a Dutch linen-draper who lived from 1632 to 1723, they opened a window on to a fascinating new world.

ANIMALCULES

Leeuwenhoek used lenses to look at cloth. But he was also a very curious man, and so he turned his lenses on other things, just to see what they really looked like, close up. Peering at rainwater one day, he was transfixed by the sight of millions of creatures darting about. He took the lens away and looked with his naked eye. Nothing. Just water. Again through the lens. And there were the creatures again, wriggling and squirming and slithering, endlessly busy.

INVISIBLY EVERYWHERE

He named them 'animalcules'. They were a human's first-ever view of what came to be called microbes or micro-organisms. (We often call the disease-causing ones 'germs'.) Leeuwenhoek began to grind and polish lenses so that he could see more clearly, mounted a lens on stand - and in the process had made his own microscope. He looked at fluid scraped from inside his mouth, skin, tree bark, leaves, the holes in rotten teeth. Everywhere he found these tiny energetic creatures, always invisible until he revealed them with his lens; he wrote excited letters to other scientists about them.

IT'S TRUE I TELL YOU! TINY ANIMALCULES EVERYWHERE!

DER LOONEY BIN

DER LOONEY BIN

BLOOD-SUCKERS

In England, Robert Hooke enthralled people by publishing a book in 1645 that vividly described the microscopic life he had seen. He watched a louse sucking blood from his hand, and wrote, 'I could plainly see a small current of blood which came from its snout and passed directly into its belly.'

MICROBE HUNTERS

Doctors of the nineteenth century became fanatical microbe-hunters. Once Louis Pasteur in France discovered that diseases and decay are caused by microbes, doctors used their microscopes to track down the germs that caused the common diseases, learning to control and kill them. But they couldn't find the germs responsible for diseases like colds, flu, measles, mumps, chickenpox and smallpox, even though they were convinced they were there.

ELECTRON MICROSCOPE

They didn't find the answer to the mystery until the 1930s, when the electron microscope was developed. This uses a stream of tiny particles called electrons instead of ordinary light, and it allows scientists to see things up to a million times smaller than the human eye can see. Then scientists saw the culprit to blame for the unexplained diseases - a microbe smaller than they had ever imagined existed: the virus.

MODERN MICROSCOPES

Now microscopes are central to almost every field of science and industry. Whenever we need to know about materials, living creatures and plants, microscopes are at the centre of the investigation. People use them to check quality in textiles, paper and printing, safety in buildings, bridges, aeroplanes and cars, to look at rock samples in mining, crops in farming, food in the preserving industry, electronics, waste, sewage and water treatment. Police use them to study fingerprints, fibres, dirt and blood. And archaeologists use them to discover how people lived in the past from traces of plants and animals clinging to buried substances.

MICROWAVE OVEN

Instant popcorn gave proof that you can cook with microwaves. In 1945 an American engineer, Percy Le Baron Spencer, put some maize in a paper bag near a radar tube. Instantly, the maize burst - popcorn. Next he tried melting chocolate. While working on radar equipment, Spencer had noticed that the

microwaves used in radar produce intense heat (*see* **Radar** *page 75*). And so was born his idea for the microwave oven. The first ovens weren't small things that can sit on a cupboard in the kitchen. Heavy and awkward, they were designed for large institutions such as hospitals and military canteens. Smaller, lighter microwave ovens for the home didn't arrive until the 1950s.

Quiz

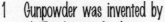

1 Gunpowder was invented by
   a) The Ancient Greeks
   b) Guy Fawkes
   c) The Ancient Chinese

2 The invention of the battery was inspired by
   a) A dead frog
   b) A dead goldfish
   c) A man whose car wouldn't start

3 Lasers are a kind of
   a) Sound
   b) Gas
   c) Light

4 What was invented in a hurry to save a president?
   a) A bullet-proof vest
   b) An X-ray machine
   c) A metal detector

5 With his microscope Anton van Leeuwenhoek discovered
   a) Microbes
   b) Animalcules
   c) Cowpox

6 When Edison was designing a light bulb he tried
   a) Hairy beards
   b) Popcorn
   c) Nylon stockings

7 The safety lift was invented by
   a) King Louis XV in France
   b) Bi Sheng in China
   c) Elisha Otis in America

8 The idea for microwave ovens came from research on
   a) A perfect toaster
   b) Radar equipment
   c) Electric stoves

9 At the beginning electricity was used for
   a) Lighting people's homes
   b) Playing games
   c) Shooting electricity through people

10 Early recipes for ink used
   a) Blood
   b) Urine
   c) Beetles

11 Benjamin Franklin is famous for
   a) Flying a kite in a storm
   b) Inventing television
   c) Making a line of monks jump up and down

12 Three hundred years ago false teeth were made from
   a) Hippo ivory
   b) Bones
   c) Rot-proof nylon

NYLON

The 'ny' in nylon comes from the initials of New York, where nylon was shown off to the public at the 1939 World Fair. It came about by accident in 1930. American scientists searching for a stronger substitute for silk prepared a mixture, but left it heating too long. When they finally rescued it, it had become a strange watery but stringy mess. Interestingly though, the fibres were so strong that they could be stretched a long way before they broke. The scientists, Julian Hill and Wallace Carothers, didn't at first realise how important their 'accident' was. But nylon quickly found important uses: immensely strong but light parachutes for aircrew during World War II; nylon toothbrushes; and fashionable, light stockings for women.

PAPER

Take old rags, mulberry tree fibres, hemp and fishing nets. Soak well in water. Crush and press. Here you have a recipe for the first paper that was good enough to write on. It was made by Cai Lun, a member of the Imperial Court of Peking, in the year AD 105. Paper-making spread all over China, Asia, and then beyond. By the fourteenth century there were paper mills all over Europe. The word paper actually comes from 'papyrus', made from the stem of a reed that was very common in the valley of the River Nile. The fibres were scraped off, twined together, pressed, then dried and used for writing on as long ago as thirty centuries BC.

PENICILLIN

KILLER GERMS

Only 50 years ago a mouth ulcer or cut on your knee could kill you. Germs - tiny living creatures invisible to the naked eye - entered people's bodies, weakened them swiftly, and could kill them. Children died from throat, stomach, brain and bone infections. Women died having babies. Doctors knew how to destroy germs outside the body (*see* **Antiseptic** *page* 9), but not once they were inside the body. All they could do was cut off the infected part to try and stop the infection spreading.

A WANDERING MOULD

But in London in 1928 a Scottish doctor, Alexander Fleming, was looking for ways to kill germs. He gathered samples of them from inside boils and noses, and grew them on a kind of jelly in small glass dishes. One day he noticed that one dish - spread with a very dangerous germ - had a mould on one edge. Around the mould *the germs had died*. He was intrigued: could the mould kill other germs? What about other moulds? He became a

fanatical mould-hunter: moulds from old cheese, boots, jam, clothes, books, he whisked them all back to his laboratory for testing. But only his first mould had the miraculous power. He named it *penicillin*. And - most important - he preserved it, and wrote down the results of his tests for other scientists to read.

BEDPAN FACTORY

Ten years later (in 1938), scientists in Oxford led by Howard Florey and Ernst Chain were looking for germ-killers that doctors could use *inside* people's bodies. They read Fleming's writings and decided to try and make enough penicillin from mould juice to test on animals and people. They needed 500 litres of juice each week for several months, to treat just five patients. Using dustbins, oil cans, baths, food tins, milk churns and library book-racks in a web of plumbing, tubes, pumps, warning lights and bells - they forged their makeshift penicillin factory. The best containers were hospital bedpans, 'sown' with mould spores using paint spray-guns.

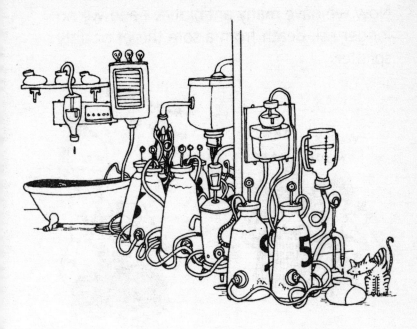

A MIRACLE DRUG AND MOULDY MARY

By 1944 the miraculous power of penicillin was clear for all to see. Doctors could now save wounded soldiers of World War II who would once have died or had limbs amputated. Fleming, Florey and Chain's penicillin was the first of the drugs we call *antibiotics*: fifteen years and thousands of world-wide mould quests passed before another strain of penicillin with such germ-killing powers was discovered. 'Mouldy Mary' (an enthusiastic mould-hunter in America) found it on a rotting cantaloup melon.

Now we have many antibiotics - and we no longer risk death from a sore throat or dirty splinter.

PENS

SPLURGE-FREE

You sometimes hear the phrase 'necessity is the mother of invention', and in the case of the fountain pen, it was probably true. As many as 25 centuries ago, people in Egypt and China wrote with reed or bamboo pens filled with ink, and 'endless pens' were found in seventeenth-century Paris. But the inventor of the first proper fountain pen was an American who had some important paperwork destroyed by a splurge of spilled ink. He was so cross that he threw himself into designing a ready-filled pen that would release ink in a free-flowing but even way. In 1884 he succeeded. His name was Lewis E. Waterman, and he perfected the system still used in fountain pens today.

BALL-POINTS

Ball-point pens were first used by pilots in World War II who had problems with fountain pens, which, at high altitudes, leaked all over maps and log books. In 1938 a Hungarian journalist, Laszlo Biro, was so impressed by quick-drying ink used in a print shop, that he developed a pen based on it. So now you know why they are called biros.

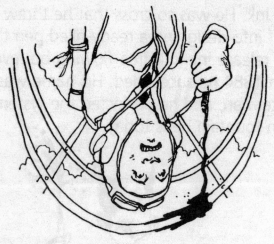

GOOSE-FEATHERS

Quill pens used to be the main writing instruments, made from a bird's wing feather, usually from geese, but also from swans, turkeys and ravens. But quill pens had to be dipped in ink, and needed constant sharpening. Dip pens with metal nibs mounted in wooden handles were developed in the mid-1700s, and continued to be used in many schools right up until the 1960s. Children's desks had their own ink-wells (usually stuffed with bits of paper and pencil-sharpenings!).

PHOTOGRAPHY

Making the first successful photograph took
eight hours of light on a summer's day. Its
creator, the Frenchman Nicéphore Niepce, called
it a 'sun drawing', and it showed the view from
his workroom window - a pigeon house, pear
tree, barn-roof and part of a house, exactly as it
looked that summer of 1826. Until then, if you
wanted to record what something looked like,
you had to draw or paint it.

PAPER PHOTOGRAPHS

These early photographs were made on large metal plates coated with chemicals; no copies could be produced. An Englishman, William Henry Fox Talbot, worked out how to create photographs on paper and make copies of them. His first pictures in 1834 were beautiful black and white patterns formed by laying plants, lace and feathers on photographic paper (paper coated with chemicals) and shining light on them. Where the light shone directly on to the paper, the chemicals made the paper turn black. But where it was covered by a solid object - the plant for example - it stayed white. Before long, in 1841, Talbot was using a camera to capture the beauty of his garden.

EARLY PHOTOGRAPHIC CAMERAS

The first photographic camera for sale was called 'daguerreotype apparatus' after the French theatrical designer, Louis Daguerre, who developed it. When he first showed his apparatus in public in 1839, people were fascinated by the idea of capturing an image of the real world - exactly as it was - to keep for ever. One Frenchman wrote that 'Everyone

wanted to record the view from his window ...
He went into ecstasies over chimneys, counted
over and over roof tiles and chimney bricks, was
astonished to see the very mortar between the
bricks - in a word, the technique was so new
that even the poorest plate gave him
indescribable joy.'

TOO BIG TO CARRY

Early cameras were large and bulky, made of
wood and needed support on a table or stand,
so they were difficult to carry about. There was
even one camera so big that it had to be
mounted on wheels and drawn by a horse.

SMALL ENOUGH TO HIDE

Yet in another 40 years cameras had become so much smaller and lighter that there was a craze for 'detective cameras' - so small that people could hide them in hats, ties, handles of walking sticks and underneath waistcoats, or disguise them to look like books, binoculars, revolvers, or even watches.

MOVING PICTURES

We didn't find out how a horse really moves its legs until 'moving' photographs. The beginnings were in 1878, when Eadweard Muybridge took pictures of a horse trotting, using twelve cameras spaced about 70 centimetres apart. As the horse passed, each camera took one picture. A year later Muybridge repeated this experiment, this time using 24 cameras positioned beside a race-track to photograph a galloping horse. From Muybridge's pictures people discovered that traditional paintings of horses galloping with all four legs off the ground - front legs stretched forward and hind legs back - were wrong. A galloping horse's feet are all off the ground together *only* when they are bunched together under the belly.

Moving pictures work because the human eye keeps an image of an object for a fraction of a second after that object is out of sight. If slightly-changed images of the object are passed in front of the eye in a rapid sequence, the eye 'sees' them as a single *moving* object. (*See also* **Magic lantern** *page 44,* **Cinema** *page 19.*)

POST-IT NOTES

A painstaking choir-singer invented those little pastel pieces of paper that stick, unstick and restick again and again. In 1970 an American scientist, Dr Spencer Sylver, discovered a kind of glue that sticks without sticking - but didn't know what to do with it. So he wrote a report, and forgot about it. A few years later, another research worker, Arthur Fry, supplied the answer. Fry was a keen choir-singer, and needed to mark his music book. But he didn't want to damage the paper. So he tried putting a layer of Sylver's strange glue on to the page markers.

To Fry's satisfaction, the markers stayed firmly stuck when he wanted them to, but he could remove them (or move them) any time he wanted. Blocks of this paper went on sale in 1979, and it wasn't long before offices were swamped with notices adorning books, papers, filing cabinets, doors - or anywhere else people cared to put them!

PRINTING

CREATED IN A DAY

The first printed book in English (in 1474) was so amazing that its printer, William Caxton, wanted all his readers to understand the miracle: he wrote in the introduction 'It is not wreton with penne and ynke as others bokes ben ... for all bokes of this storye were begonne in oon day and alsy fynisshid in oon day.'

THE BEGINNINGS

Imagine that each page of each book has to be made individually by hand. It would take months. There wouldn't be many books around, and few people would ever get to read them (or learn to read in the first place). This is what it was like before the era of mass printing by *movable type* which began in the fifteenth century.

Printing was originally a kind of 'stamp pad' process: a separate 'printing block' was made for each page of a book by carving all the words and punctuation on a wooden block and then inking it. The ancient Chinese, inventors of

so many things including paper (see **Paper** *page 56)*, also developed this kind of printing. As long ago as AD 600-609, they created whole pages of printed writing. There was a printed newspaper in China in AD 748. The earliest complete printed book (actually a paper scroll) was made in China in AD 868.

I want to change the will again. Get your chisel out Charlie.

MOVABLE TYPE

It was also a Chinese printer, named Bi Sheng, who invented, around the beginning of the eleventh century, a form of type made from baked clay. This was 'movable' because a piece of type was made for each of the characters of Chinese writing; they were placed together to

make words, used to
print, then broken up
and placed in a
different way for a
different word. In 1297
another Chinese
printer, Wang Zhen,
made separate
characters out of hard
wood. He had 60,000
characters and used
them to print a local
newspaper: producing
100 copies took him a month.

SPREADING TO EUROPE

We don't know exactly how printing spread to
Europe. But we do know that in the 1420s and
1430s a Dutchman named Lauren Janzoon
Koster tried to fashion movable type in the Latin
alphabet (the one we use). But the letters are
much smaller than Chinese characters, so he
had great difficulties and never really solved the
problem. Other craftsmen started using single
letters made from metal, pressed into clay to
form a mould for a page. These could be
re-used on a new slab of clay for another page.

They then poured molten lead into these moulds to make the printing blocks. But it was still very slow, and there were still few books around, each very expensive.

MASS PRINTING WITH MOVABLE TYPE

Imagine, then, the wonder when someone invented a way to make many copies of each book, very fast! The first book of this kind was a Bible, printed in Mainz (Germany) in 1455 by Johannes Gutenberg. He was a goldsmith skilled at working in metal. We don't know for certain, but he probably developed his method without knowing what was happening elsewhere, as often happens (*see* **Turning points and giant leaps** *page 25*).

GUTENBERG'S INVENTION

First he invented a printing press, developed from a machine used to press grapes for wine-making. Then he made individual metal stamps for each letter. He bolted these together in a frame to form the printing plate for a whole page of words. But, most important, when he finished printing that page, he could take the letters apart easily and reassemble them to

make the next page. If you've ever tried to print again and again using a stamp pad, you'll realise that he had to solve a lot of problems to make it work well. In particular, the type had to make a strong, regular print across the whole page, and not wear out. Every letter had to be the right size to make the same strength of mark, the ink had to go on smoothly, and the printing machine had to press down evenly across every page of the whole book.

A REVOLUTION

Knowledge of this kind of printing spread to other countries - and it changed the world. Books were much cheaper and plentiful. People were able to read more, learn more, and spread their knowledge and ideas rapidly. More books were made in the first 50 years following Gutenberg's invention (up to the beginning of the sixteenth century) than had been produced in the whole of the previous 1000 years.

Electronics and All That

'Electronics' is transforming our world even as you read this. Electronic devices do not just *use* a flow of electricity. They are able to control how it behaves, manipulating the tiny particles called *electrons* in an exact way to make the flow of electricity extremely efficient. This is where the name 'electronics' comes from.

The first electronic devices were in radios in the 1920s and 1930s and were called *valves* or *vacuum tubes*. They looked rather like light bulbs, were fragile, and got very hot: one computer had 18,000 and was as big as a large room.

But then came the *transistor* — a tiny piece of crystal that scientists discovered could do the same work as a vacuum tube. And so the era of portable electronic equipment began with the first transistor radios in 1955.

It was just the beginning: scientists found they could pack more and more transistors on to each crystal. Now there can be millions on a *chip* only 1 millimetre square, serving as a central processing system of a computer.

What would once have filled a very large hall can now fit into a very small bag — and can do much, much more.

RADAR

ALL DOWN TO CARROTS?

During World War II radar stations around the coast of Britain helped warn that enemy aircraft (invisible to the eye) were approaching. British fighter-planes, equipped with radar, were able to find the planes and intercept them, flying more safely at night and in poor conditions. Ships could detect enemy submarines and convoys at a distance. But *how* they did this was all kept top secret. Very few people knew anything about it until after the war. Instead people were told that British nightfighter pilots just ate lots of carrots to improve their night vision!

MICROWAVES AND ECHOES

Radar actually stands for *RA*dio *D*etection *A*nd *R*anging, and it is a way of locating far distant objects using a very small radio wave called a microwave. Microwaves are sent out: they bounce off a distant object. And from the time it takes for the echo of the radio wave to return, (and the direction it comes from), it is possible to work out where the object is, and how far away. Scientists in many different countries, particularly Britain and America, worked for many years to develop radar. But it was not easy. Radar only works if the beam of radio waves sent out is very strongly focused. Otherwise, the echo thrown back from the water or the earth may overwhelm the much weaker echo from the target.

SMALLER AND STRONGER

In 1935 an Englishman, Robert Watson-Watt, found a way to amplify the radio waves (make them stronger) and focus them very tightly. With equipment set up in the back of his van, he detected objects 64 kilometres (40 miles) away. Other British scientists, John Randall and Harry Boot, made further important developments in

1940. These made it possible to send much, much smaller microwaves, even more powerful and highly-focused, using equipment small enough to install in ships and aircraft. At the beginning of 1940 radar sets were as big as garages. Five years later they were as small as suitcases. (*See also* **Microwave oven** *page 51.*)

There's nothing down there.

RADIO

Guglielmo Marconi was so excited when he first sent radio signals from one end of a room to the other that he dragged his mother out of bed in the middle of night and insisted that she come and listen. That was just over 100 years

ago, in 1894. Marconi was only twenty, working in dusty attic rooms in his home in Italy. For months he had struggled to make the radio waves travel further than just a few metres, fiddling with batteries and bits of copper wire, sheets of metal from old water tanks, metal balls, tubes of metal filings - trying to get his makeshift equipment to work better.

MAMA! come See! I've invented da radio!

that's a nice Guglie Give me a shout when "Woman's Hour" is on.

CAN'T BE DONE

Scientists were certain that he would not be able to make the radio waves travel very far. They had been working for some years to find out about these newly-discovered, still mysterious, invisible electrical waves. First (they said), he wouldn't be able to make the radio waves strong enough to travel more than a few kilometres at the most. Second, radio waves moved in straight lines. So, as the earth was curved, radio waves travelling any distance would just shoot into space.

A VISION

Marconi knew all this, but he had a vision of 'wireless' - communicating across the world without needing electrical wires linking sender to receiver (*see* **Telegraph** *page 101*). No matter what anyone said, he was determined to make it happen. Within six years (in December of 1901), he made the jump from the attic in his home to winging wireless (radio) waves across the world.

Twenty years later scientists discovered why he succeeded. Although radio waves do shoot

away from the earth's surface, they are bounced back again by a layer of tiny electrified particles high up in the atmosphere. But in December 1901 people only knew that, in the twinkling of an eye, radio signals had sped 3000 kilometres across the Atlantic Ocean between England and Canada - just as Marconi said they would.

LIFE-SAVER

The life-saving power of Marconi's radio was shown to people in the most tragic way one freezing night in 1912. On 15 April, just after midnight, the largest and most luxurious ship then built - the *Titanic* - was steaming through the Atlantic on her first voyage. Everyone believed she was unsinkable. At 11.40 pm she struck an iceberg. Two hours forty minutes later, she slid beneath the black, freezing waters of the North Atlantic, taking 1500 people with her. The only survivors were the 700 people who managed to cling on to life in the few life-boats, and were saved by the *Titanic*'s radio distress calls. These were heard by another ship, the *Carpathia*. She arrived two hours after the black waters had swallowed the great liner, but soon enough for those in the lifeboats to be still alive.

TRANSFORMING THE WORLD

For anyone at sea, in the air, or in any place where wires cannot link them to others by telegraph or telephone, radio is a life-line. Now, it can penetrate even the ocean depths and outer space. Entertainment broadcasting (beginning in the 1920s), and the kind of radio we have now, with voices and music from anywhere in the world at the push of a button or turn of a knob, was still to be developed by others who came after Marconi. But his wireless signals across the Atlantic were the vital beginnings.

RADIUM

Inventions can sometimes be very dangerous, because at first people do not understand them fully. In 1898, the French scientist Marie Curie discovered a new substance, which she called radium. Doctors soon found it could be used to treat cancer. Radium saved and lengthened millions of lives. But the radiation from radium also *causes* cancer: we know now that people must protect themselves from it. At the beginning people were unaware, and when

radium's healing powers were first revealed to the public, 'radium inventions' became all the rage.

DANGEROUS GIMMICKS

Silly and dangerous radium gimmicks went on sale: radium Vita face powder and beauty creams, radon gas 'sparklets' for making soda water, fake medicines sold by dishonest people pretending to be doctors. All of these put large doses of radiation into people's bodies. There were also workers painting luminous numbers on clocks and watches. They licked their paint brushes to keep the tips fine, and each lick gave them radiation from the radium-based paint. Many died from jaw cancer. By the time all these horrors were understood, it was too late: Marie Curie herself died from years of radium radiation, and thousands of laboratory workers all over Europe and America showed the appalling effects.

RECORDING

A machine captured the sound of someone's voice for the first time on a cold December evening in 1877. People felt their hair prickle, it was such a weird, ghostly sound - a voice reaching them *from the past*. Though the first words recorded were not at all eerie:

Mary had a little lamb,
Its fleece was white as snow ...

When the American Thomas Alva Edison showed his 'talking machine' to journalists, they flew into raptures, wrote excited newspaper reports, and soon had people flocking to hear the contraption perform its magical tricks.

THE PHONOGRAPH: GROOVES AND VIBRATIONS

That first machine was called a *phonograph*, and also a 'sound writer'. This is how it worked: Edison turned a handle to turn a cylinder covered with tinfoil. He spoke into the machine. The sound of his voice made a diaphragm (a thin plate) vibrate; this was fastened to a needle that also vibrated, pressing a groove in the tinfoil round the turning cylinder. To play the

words back, a blunt needle moved across the grooved tinfoil, causing the linked diaphragm to vibrate. It gave out the sound again - scratchy and faint - of the words said earlier. A later version of the phonograph had *wax* cylinders and boosted the sound with a horn. (*See also other Edison inventions:* **Light bulb** *page 42*, **Child inventors** *page 123*).

THE GRAMOPHONE: THE FLAT RECORD

So began the world's sound-recording industry! Edison's phonograph was the ancestor of the gramophone and the record player. The gramophone, which plays flat records with grooves (instead of cylinders), was invented in America by Emile Berliner in 1887. But budding pop-stars had an exhausting career ahead of them: it was another twelve years before Berliner developed a mould for making many copies of the same record. Until then, performers had to record the music again and again to make enough records to sell!

C'mon boys... just one more and we'll call it a day.

THE TAPE: USING MAGNETISM

The first 'tape' recording was actually made on wire. It was invented in 1898 by a twenty-year-old Dane, Valdemar Poulsen, but did not become widespread until 30 years later, from 1927 onwards, using steel tapes on big, heavy reels. Ribbon-tapes coated with a metallic substance came later, in 1935 - at first on large reels. The tapes weren't easy to use: they liked tangling, kinking and snapping, or winding themselves off the reel into a bird's nest on the floor. The neat, compact sealed cassette tapes that we can buy now first arrived in the 1960s.

Tape recording works differently from the groove arrangement of records. The recording microphone sends out electricity in a pattern of strong and weak pulses to match the loud and faint sounds. These electric pulses are stored as *magnetism* on the surface of the tape. The sounds are replayed when the tape passes a magnetic head that interprets the varying strengths of the magnetism as different sounds.

THE CD: USING LASERS

Compact discs (CDs) are different again, born of the electronic age. They have billions (thousands of millions) of 'pits and flats' (bumps and holes) on each disc, so small they can only be seen with an electron microscope. Laser light scans the disc; the reflection flickers (at over a million times each second) as it passes over the pits and flats. Electronic detectors pick up the flickers and turn them into electrical signals, transformed back into the vibrations of voice or music. (*See* **Electronics and all that** *page 74 and* **Laser** *page 39.*) CDs sprang from a partnership in 1979 between the Dutch company Philips and the Japanese company Sony.

REFRIGERATOR

ELECTRIC FRIDGES IN THE HOME

Leaking gas from the first electric fridges stank so much that the cooling machinery had to be put in a separate room. It wasn't until the 1930s (over twenty years after the first home refrigerator went on sale in America in 1913) that you could get cold food without a stink. By then the machinery was sealed into an airtight casing.

In Britain now we take it for granted that homes have fridges. But right up to the early 1960s, they were luxuries. Many people still used larders and cellars, shopped daily and kept milk in cold water or on window-sills in winter.

ICE CUPBOARDS

At the very beginning of the twentieth century some homes did have 'refrigerators'. But these were large slate-lined cupboards kept cold by enormous lumps of ice. Cut from frozen lakes in winter or made in factories, they were kept in ice stores and delivered daily by 'ice men'.

ROBOTS

Human-like robots with flesh, bones and internal organs are no longer just the stuff of science fiction. Some breathe, blink, have beating hearts, human temperature and skin with the strength - and weakness - of human flesh. You can reach their insides through a zip. Computer-controlled, reacting to drugs and medicines, they are used to teach medical students how to deal with emergencies - such as a patient's heart stopping.

Special robots also test car safety, plane ejector seats and space equipment. One for testing military clothing actually sweats and can crawl about on all fours. There are police warden-robots, security guards, robot-sheepshearers (shearing a sheep in 100 seconds instead of three minutes taken by a human) and radio-controlled robot-jockeys for exercising horses. Some are designed to go where people can't - into damaged nuclear power plants when leaking radiation would kill a human being.

Ever since the 1960s, vast numbers of robots are used in industry. But these usually look just like machines, with gripping devices at the end of

mechanical arms. They can do regular, planned tasks - assembling cars, washing machines and televisions. (*See also* **Automaton** *page 11.*)

Quiz

1 The first successful photograph in 1826 was made on
   a) Paper
   b) The lid of a hat box
   c) Metal

2 Penicillin is made from
   a) Old cheese
   b) A mould
   c) Papyrus

3 What discovery led to dangerous gimmicks on sale?
   a) Antiseptics
   b) Radium
   c) Antibiotics

4 Radar is a kind of
   a) Mexican wave
   b) Tidal wave
   c) Radio wave

5 The inventor of the first movable type was
   a) Johannes Gutenberg
   b) Bi Sheng
   c) Louis Braille

6 The first recording machine used
   a) Metal discs
   b) Tinfoil
   c) Paper ribbons

7   Newly-invented nylon was used for
    a) Telegraph cables
    b) Post-it notes
    c) Toothbrushes

8   Radio signals first crossed the Atlantic in
    a) 1741
    b) 1901
    c) 1941

9   What was used for writing before fountain pens?
    a) Biros
    b) Typewriters that looked like squirrels
    c) Feathers

10 Robots have been invented to
    a) Shear sheep
    b) Have heart attacks
    c) Ride horses

11 Mass printing with movable type began in
    a) The fifteenth century
    b) The fifth century
    c) The twentieth century

12 Post-it notes were invented by
    a) A Scottish dancer
    b) A Chinese painter
    c) An American choir-singer

SEWING MACHINE

Tailors burned the first really practical sewing machine and nearly killed the inventor, Barthélemy Thimmonier. They were terrified it would put them out of work. That was in France in 1830. Fifteen years later, in the United States, Elias Howe developed the first successful sewing machine. It was different from the ones we have now - it held the cloth upright while the needle moved sideways. This, and others that had gone before, were for factories. In 1851, also in America, Isaac Singer designed one for home use, the ancestor of the kind we have now. He was brilliant at demonstrating his machines, even at circuses, and they became immensely fashionable - preferably disguised as lions and squirrels!

Once, all clothing and leather goods had to be made by hand. With newfangled sewing machines, clothes could be turned out fast and in great numbers. A huge industry sprang up to produce cheap clothes, machined boots and shoes, horse saddles and harnesses, umbrellas and mattresses - things that ordinary people, not just the rich, could now afford to buy. In its own way, the little sewing machine has helped transform the world!

SPORTS

BASKETBALL

Most sports, like football and cricket, weren't 'invented', but developed slowly over hundreds or even thousands of years. By contrast, basketball was created - very quickly in December 1891 - with the peach baskets as goals. Students at an American YMCA training school were bored with PE classes, and James Naismith looked for something to inspire them. He mixed ideas from hockey, American football and soccer, blended them with his own, and came up with basketball. The sport was instantly popular and spread to Canada, Europe and the East. The iron hoop and net came two years later. But players had to climb a ladder to retrieve the ball - until someone had the sense to cut a hole and let the ball drop through!

it's a great game but it's costing us a fortune in balls...

GOLF

The first golf balls (in Scotland in the fifteenth century) were stuffed with feathers. The Scots enjoyed the new game so much that in 1457 King James II of Scotland banned it: his subjects were wasting too much time! But the game remained popular, and Mary Queen of Scots, famous for having her head chopped off by Queen Elizabeth I, also has the distinction of being the first woman to play the game.

RUGBY

In England a pupil at Rugby School broke the rules of football, thus inventing rugby. In 1823 he grabbed the ball, clutched it to his chest, and streaked off towards the opponents' goal. A new game using hands developed fast, though the rules were vague until 1871 when the Rugby Union and the fifteen-player game began.

SKIING, SKATES AND SKATEBOARDS

The first *ice skates* were bone, and used at least 2000 years ago. We know of them in the first century AD in Scandinavia, but they may be even older. *Skis* have also been used for travel in snowy regions for thousands of years (although the ski with the curved front is much more recent, dating from 1880). *Roller skates* are newer than skis, though they still date from about 200 years ago, called ground skates. They were devised in the eighteenth century by an unknown Dutchman, and were a four-wheeled rigid sole tied to the shoe.

There is a story about the first known roller-skate maker - a Belgian called Joseph Merlin,

who had a shop in London around 1760. He
wanted to impress everyone by gliding stylishly
into a grand ball, playing the violin. Unhappily,
he hadn't worked out how to stop. He slid on
helplessly, the slippery length of the entrance
hall and - to the dismay of the spectators -
straight into an expensive mirror. Mirror, violin
and Merlin were all horribly smashed.

There goes Jo.
anything for a laugh.

Fortunately, Merlin lived to tell the tale.
It wasn't easy to go round corners with these
early skates but they became more popular once
skaters weren't rattled to bits on bumpy roads:
shock-absorbers were added by American James
L. Plimpton in 1863, and in 1884 ball-bearings in
the wheels, (the idea of another American,
Levant Richardson).

Skateboarding mixed surfing and roller skates, as shown by its first name - 'rollsurf'. Two Californian surfers, Mickey Munoz and Phil Edwards, dreamt up the idea in the 1960s for practising surfing. It didn't catch on in its own right until the 1970s.

SKITTLES AND TENPIN BOWLING

Knocking down the devil was a favourite game in medieval Germany - or rather knocking down a skittle to symbolise him! Some forms of the game are very old and appeared in stories about Ancient Greece. Nowadays the game is played in different ways all over the world. Tenpin bowling is a 'modern' form, though it began over 100 years ago, in 1874.

Johann! Johann! Let us play knock down the devil again!

STEAM ENGINE

When inventors first try out ideas, they have to make use of whatever bits and pieces suit their purpose - however bizarre. In 1765 the Scottish instrument maker, James Watt, was trying to find a way to make the steam engine more powerful and efficient. He needed a large cylinder and piston to test his plan. The largest he could find was an anatomist's syringe for squirting wax into dead bodies before cutting them up for studying. Undaunted, Watt used it, and thus he made the first working model of an engine that would transform the world for all time.

Before this, machines could be driven only by animal or human muscle, or by wind- or water-power. The invention of the working steam engine in the early 1700s (for pumping water out of coal and tin mines) began to change all this (see **Turning points and giant leaps** *page 25*). In particular, James Watt's improvements (anatomist's syringe and all) led to a faster and much more powerful engine. Fuelled by burning coal (sometimes wood), people harnessed it to drive all kinds of machinery in industry, beginning with the

cotton mills in 1785. Then came the revolution in transport: the steamboats in the 1780s and railway locomotives in the 1800s.

it's a kind of furry feeling...

MIAOW

STETHOSCOPE

You can hear lungs working inside someone's body if you hold a roll of paper to their chest. René Laennec, a doctor in Paris, discovered this - possibly because he was squeamish about putting his ear to a patient's body. And, with his simple roll of paper, he had invented the stethoscope. In 1819 he began using one made of wood. By the 1850s doctors found it was easier to use two tubes - one for each ear -

joined to a single piece placed against the chest. And so we have the beginnings of the modern stethoscope ...

TELEGRAPH

SENSATIONAL CAPTURE

A murderer tracked from his crime - and caught - dramatically proved the value of the new electric telegraph. Just over 150 years ago in London, on 1 January 1845, the telegraph operator at Paddington railway station received a telegram:

> *'a murder has just been committed at salthill and the suspected murderer was seen to take a first class ticket to london by the train which left slough at 7.42 pm he is in the garb of a kwaker with a brown greatcoat which reaches nearly down to his feet he is in the last compartment of the second first class carriage.'*

John Tawell had killed his girlfriend. Police tracked him by horse-bus from Paddington and arrested him. He was tried, convicted and executed, and Londoners aptly said 'The cords (telegraph wires) have hung John Tawell.'

SIGNALLING WITH ELECTRICITY

Before the electric telegraph, messages had to be *physically carried*, by runners on foot or riders on horse-back, by carrier-pigeons, by ships or trains. People could send very brief fire-beacon or flag signals from towers and hills, or between ships at sea, *as long as people were close enough to see them*. Distant signalling - out of sight-range - was impossible. In several countries scientists experimented with electricity to send messages along a wire from place to place. But a general long-distance system wasn't designed until the 1830s, in England, by William Fothergill Cooke and Charles Wheatstone. These first electric telegraphs moved magnetic needles to point to letters of the alphabet on a dial.

MORSE CODE

Later, Morse code was used - so called because it was invented in 1837 by the American Samuel Morse. He converted each alphabet letter into a pattern of dots and dashes. To send a message, the telegraph operator pressed a small lever to make an electric current flow along a wire. Depending on how often he pressed the lever

and how long he held it down, he could make long and short electrical flows to match the dots and dashes in a message. At the other end, these electric pulses were changed into long and short clicks or buzzer sounds, or printed on a 'Morse writer' as dots and dashes.

HUMMING WIRES

Some people were more suspicious than happy at the humming telegraph wires. When Samuel Morse built the telegraph in America, enraged Kentucky farmers tore down the first poles and wires, convinced they were spoiling the harvest.

ACROSS THE SEAS

A puzzled fisherman managed to thwart the first efforts to lay the telegraph under water. In 1850 he was fishing in the English Channel and found his anchor snarled up with a strange, thick rope with a copper core. So he cut it - and so slashed the first-ever telegraph link between Britain and France, laid with much fanfare only the day before! It took a whole year to organise a replacement. Forty kilometres (25 miles) long, four copper wires inside gutta percha (a rubber-like substance), weighing about 180

tonnes, this time it was armoured with iron wires and worked for eight years before needing big repairs.

GIANT NETWORKS

By the 1890s telegraph wires criss-crossed Europe and the United States - two great networks joined by the massive Atlantic cable. Even so, it took ten hours to send a message to Australia in 1872. About twelve telegraph stations had to relay the message - read it and retransmit it onward. Mistakes could creep in at each stage. No wonder some inventors dreamed of communicating instantly, through the air, without electric wires that need laying and keeping in good repair. (see **Radio** page 77; see also **Telephone** page 105).

TELEPHONE

The first 'telephone' receiver was fashioned from a violin case, a hollowed-out beer-barrel bung and a stretched sausage skin. Listeners at the first demonstration on 26 October 1861 believed they heard a song sung in another building. What they actually heard were just squeaks and trills, not recognisable words. But it was nevertheless *the sound of the human voice*, being transmitted down electrical wires. This was the work of a German named Philipp Reis, the first to name the contraption a 'telephone'.

TRIALS WITH A DEAD MAN'S EAR

Fifteen years later, a young Scottish American named Alexander Graham Bell performed some gruesome experiments with a dead man's ear. (He had worked with deaf people and knew a lot about how sounds travelled.) He arranged a fragment of the dead man's ear and bones so that a straw touched the eardrum at one end and some smoked glass at the other.

When Bell spoke, the vibrations in the eardrum made the straw move and scratch tiny markings on the glass. He was able to see exactly how

vibrations pass through the thin membrane of a real eardrum, and so he developed a flat, thin disk of iron - called a diaphragm - that could vibrate in the same way. He managed to convert the vibrations into an electrical current along a wire, then changed them back to sound at the other end.

THE FIRST MESSAGE

On a March morning in 1876, Bell sent the first-ever telephone message to his assistant, Thomas Watson. The telephone only worked one way, so Watson couldn't reply. But the delighted Bell wrote to his father, 'I have at last struck the solution of a great problem and the day is coming when telegraph wires will be laid on to houses just like water or gas - and friends converse with each other without leaving home.'

TELEVISION

STRANGE CONTRAPTIONS

The first television seen in public alarmed people because it showed events happening in another room, and newspapers wondered if it could see through brick walls, invade people's lives, or even see into their souls. That was in London in 1926, and the television was constructed from wood and card. It was invented by John Logie Baird, and his first working model, only a year before, was even stranger - a knitting needle, the lid of a hatbox, an electric fan motor, and torch batteries, all put together on top of an old tea-chest. This contraption managed to send the picture of a cross over a distance of three metres.

DIM AND FLICKERING BEGINNINGS

Yet within four years the BBC made its first public television broadcast. The announcer had to be fit and energetic - speak first into a microphone, then leap in front of the camera! And the first television plays showed only the actors' heads - one at a time. You could watch these blurred, flickering images on a grand total of 30 television sets in Britain.

ELECTRONIC DEVELOPMENTS

Baird's invention was different from the television system we have today. His 'camera' used two discs that revolved rapidly so that a sequence of holes in them scanned a brightly-lit scene. The light passing through the holes was converted into an electric current and sent to a receiver that had a similar disc-light system. In the late 1920s Vladimir Zworykin in America, and (though few people in Europe know this) Kenjiro Takayanagi in Japan, both developed a way to get the same result using *electronic scanning*. Electronic television soon replaced the mechanical system (*see* **Electronics and all that** *page 74.*)

Ah John, you're always infront of that thing will you not go out and play in the park?

TINNED FOOD

You can thank a French sweet-maker for tinned food. He was the first to have the idea, in the 1780s, that heating sealed jars of food could stop food going bad. It was a remarkable thought, because no one knew then that food went bad *because it had germs in it*, or that heating the germs killed them. (It was another 70 years before another Frenchman, Louis Pasteur, began to explain this.) The French sweet-maker, Nicolas Appert, just found that his idea worked. In 1795 he won a prize for it from the Emperor Napoleon, and began selling foods in sealed bottles, calling it 'appertisation'.

Fifteen years later the 'preserved food' industry spread to England and beyond - first with tin-lined containers, then with metal boxes. Unfortunately, you needed a hammer and chisel to open them! Tin-openers - just a lever with a spiked cutter on the end - didn't appear until the 1850s, 40 years later. Another 20 years passed before, in the 1870s, someone invented the twisting type with rolling cutters that we use now.

TOASTER

Early pop-up toasters had a nasty habit of shooting flaming toast into the air - thanks to a too-strong spring and a crude timing mechanism. They went on sale in 1919 in America. Ordinary electric toasters - in the beginning just bare wires that heated - began in the 1890s and only the rich could afford them. Those toasted just one side of the bread at a time, and you had to watch carefully or they caught fire. Next came fancier styles with an opening side hatch that you could pull down. This let the toast slide down so that when you shut the hatch, the toast had turned over - without anyone touching it.

I shall have cereal tomorrow.

TOOLS

TRIED AND TESTED

There was a court in Ancient Greece where
tools were brought to trial before a judge -
charged with injuring their users! Obviously no
one knew the saying 'It's a bad workman that
blames his tools.'

THE FIRST TOOLS

Humans began producing the first tools about
2.5 million years ago. They were sharp-edged
implements, made by hitting two stones
together to make a cutting edge or break off a
sharp flake. Modern scientists have tried making
some - and it's not at all easy. Early 'primitive'

humans had quite a skill, knowing which stones were suitable, how hard to strike them, and at what angle. The earliest versions were very small flakes (about 2.5 centimetres long and surprisingly sharp, for cutting meat, wood and plants), and hammerstones, choppers and scrapers also shaped by knocking flakes off a larger chunk of stone. In 1959 Louis and Mary Leakey were the first archaeologists to find these - at Olduvai Gorge in East Africa.

Archaeologists have also found larger, more complicated stone hand-axes, cleavers and picks that humans began to make about 1.4 million years ago. Again, modern scientists needed months to develop the skills to make any nearly as good as the originals.

TOOTHBRUSH AND TOOTHPASTE

Stranded without your toothbrush? Never despair! Just grab a fistful of couch grass and brush away. In ancient times, this is exactly what toothbrushes were made from - and in remote parts of the world plant fibres still do the job. Toothpastes in ancient times were also curious -

chalk or crushed charcoal mixed with juices, pastes, crushed herbs and - human urine! (Urine has ammonia in it, discovered again as a useful ingredient by modern toothpaste makers.) A recognisable toothbrush first appears in a fifteenth-century Chinese painting, and it seems to have taken another 200 years before anything like it turned up in Europe. Toothbrushes were one of the first uses for nylon in America in 1938 - known as Dr West's Miracle Toothbrushes (*see* **Nylon** *page 55*).

TYPEWRITER

The newly-invented typewriter was called a 'literary piano' and produced a peculiar hostility: doctors accused it of causing tuberculosis and bringing on madness; Queen Victoria exploded in fury when someone gave her a 'chirographed' (typed) paper. Early machines in the 1800s were actually slower than hand-writing. The first practical typewriter was made by Christopher Latham Sholes in 1867. He went on to sell an improved model, surprisingly, to gunsmiths called E. Remington and Sons. These weapon-makers began to make and sell typewriters in 1874.

The Remington was heavy and difficult to handle, and for a long time the writer couldn't see the text as it was typed! This was solved in 1890; after that, typewriters became popular. Then electrical models developed and both kinds reigned unchallenged - until electronic typewriters with small memories came along in the 1970s and then word-processing with computers in the 1980s.

VACCINATION

BLISTERS ON A MILKMAID'S HAND

A young boy bravely risked his life in a very daring experiment 200 years ago. A doctor scratched the boy's arm, took liquid from a blister on a milkmaid's hand and put it in the scratch. Then the doctor, Edward Jenner, and the boy, James Phipps, waited to see if James would fall ill.

In those days nearly everyone caught *smallpox*, a disease that killed one in every twelve of its victims. Survivors were smothered head-to-foot in hideous scars, and often blinded. Doctors couldn't treat the disease, but they *did* know that once someone had it and survived, they

never got it again. Some tried giving tiny doses of smallpox by injecting the liquid from a blister. If it worked, the patient got smallpox mildly and that stopped them catching it again. But too often this method -'inoculation' - went wrong. People died.

Which of you two hasn't had the cowpox injection?

COWPOX AGAINST SMALLPOX

Dr Jenner had heard country people say that if you had *cowpox*, a cattle disease, you never got smallpox. In 1796 James's mother was terrified that her son might catch smallpox and asked Dr Jenner to inoculate him. Jenner agreed - but

using *cowpox*. James's arm came up in nasty, weeping blisters. But then it healed, and he remained well. Two months later came the dangerous test: Jenner gave James a strong dose of *smallpox*. To the delight and relief of everybody (not least James), the boy stayed healthy.

SMALLPOX BEATEN

All over Europe doctors began to copy Jenner. In the 100 years from 1700 to 1800, smallpox had killed 60 million people in Europe. Following Jenner, deaths each year in England alone fell to less than a third. (Nowadays the disease has been wiped off the earth: the last case was in Somalia in Africa in 1977, more than twenty years ago.)

FIGHTING STRONG GERMS WITH WEAKENED GERMS

Eighty years later than Jenner's experiment, in 1878, the French scientist, Louis Pasteur proved that injecting the *weakened* germs of a dangerous disease can safely protect someone from that disease. It forces the body to develop weapons, so that it is ready to defend itself.

Pasteur named this 'vaccination' in honour of Jenner: another name for cowpox was *vaccinia*, from the Latin word for cow, *vacca*.

VACUUM CLEANER

Fashionable dinner parties at the beginning of this century might offer a strange evening's entertainment - watching the electric vacuum cleaner at work! The new gadget was so remarkable (and so expensive) that people were enormously keen to show it off. Salesmen bribed customers with a free (dusty) rug so that collected filth could be proudly displayed in a heap.

NOISY SERPENTS

The first machine was too big to take inside a house. Parked outside on a large horse-cart, long hoses were poked through open windows to suck out the dust. Hence its nickname - the 'noisy serpent'. It was invented in England in 1901 by Hubert Cecil Booth, a bridge engineer, who then started a cleaning service in London. Early small domestic vacuum cleaners, before electricity, were pumped by foot, some needing two people to work a double bellows.

VELCRO

In 1948 a sharp-eyed Swiss engineer named
Georges de Mestral noticed something curious
about burdock seeds. They clung firmly to his
clothes and his dog, and were *extraordinarily
difficult* to brush off. Intrigued, he looked at
them under a microscope and saw that tiny
hooks round each seed head gave it the
powerful grip. And then he had his brilliant idea
- similar hooks on fabric might cling together as
a fastener! It wasn't so easy - he took eight
years of trial and error to get it right. A nylon
strip with thousands of small hooks, another

with smaller loops: press the two together - you have a fastener that closes quickly and firmly. The name Velcro comes from the French words *velours* (velvet) and *crochet* (hook).

WASHING MACHINE

Washing machines started life as wooden barrels that were turned or rocked with a handle to tumble the washing about. When electricity was added, doing the laundry became a little risky! The first electric machine, designed by A.J. Fisher in 1908, had the motor under the wooden washing tub: not surprisingly, it tended to get wet and give dangerous electric shocks. Even so, it was expensive - for the rich only. All the early machines had mangles - two rollers turned by a handle. You fed the laundry through to wring out the water. Spin-driers became popular in the 1960s. But the principle of washing with a revolving drum to slosh clothes about in soapy water hasn't changed since the beginning. Luckily, though, machines are now safer.

X-RAY

In December 1895, a German professor, Wilhelm Konrad Röntgen, had an early Christmas present - a picture of the bones in his wife's hand. He had discovered a kind of ray that passed through human flesh but not bones, and so could make a shadow-picture of a skeleton inside a living body. He called the strange rays 'X-rays', because they were unknown. But news of his discovery appalled some people. Seeing inside people was a threat to privacy! One professor was so horrified at an X-ray of his own skull that he couldn't sleep a wink afterwards.

EXTRAORDINARY USES

Enthusiasm soon outweighed fear, and extraordinary uses were dreamed up: one man claimed he could photograph the human soul and showed 400 photos to prove it! Within the year though, serious uses began: doctors took successful X-ray pictures of a needle in a dancer's foot, a human foetus inside its mother, and the skull of an Egyptian mummy. And one even watched pearl buttons pass through the insides of a dog.

LITTLE CURIES

In World War I X-rays saved many lives, locating bomb fragments and bullets in the flesh of wounded soldiers, and showing clearly how bones were fractured. Marie Curie, discoverer of radium (*see page 81*), designed the first mobile X-ray unit. It was just an ordinary van with an X-ray machine, staffed by a small team including her seventeen-year-old daughter. But it could travel right to the battle-front. In just two years more than a million men passed through the twenty mobile cars (known as Little Curies) and 200 hospital X-ray units that Marie organised.

A THOUSAND USES

Nowadays, X-rays are also used to look at fractures in industrial machines and pipes, packages in luggage, and a thousand other uses where people need to see through wood, metal, or live flesh. Combining X-rays with computers now allows doctors to make even more complicated and detailed scanning investigations by building up three-dimensional X-ray images.

Child Inventors

Have you ever dreamed up message and lift systems for your house, transformed your bike into a rocket or built a car out of cardboard boxes? Then you're in good company with famous inventors who started inventing as children: remember *Louis Braille* and *Charles Babbage*? (See pages 13 and 22).

Guglielmo Marconi was also forever concocting some gadget. He transformed a sewing machine into a roasting-spit, erected a spear-like contraption on the roof to catch electricity from a storm and ring bells inside the house, smashed dinner plates in elaborate experiments, and altogether made his father furious. But Marconi was already gripped by a fascination that would lead to the invention of radio.

Thomas Edison's mother gave him a book of science experiments when he was nine. He did them all, then made up his own – filling his house with thuds and explosions (to the horror of his mother). As a twelve-year-old, in his first job, he installed a mobile scientific laboratory in an empty luggage compartment on a train. That is, until he started a fire by mistake and the guard threw everything out.

So if you like designing and making things . . . who knows? Perhaps one day *your* name will be in a future book about inventions . . . (See pages 42, 77, 83).

Quiz

1 The first telegraph laid under water was damaged by
   a) Angry Kentucky farmers
   b) A puzzled fisherman
   c) Angry killer whales

2 The first stethoscope was
   a) A wine bottle
   b) A roll of paper
   c) A tin can

3 The first television picture showed
   a) A UFO
   b) A cross
   c) An out-of-breath BBC presenter

4 Tinned food was invented by
   a) A Scottish fisherman
   b) A French sweet-maker
   c) An Italian scientist

5 The first basketball goals were
   a) Shopping baskets
   b) Peach baskets
   c) Laundry baskets

6 The first practical sewing machine was invented in
   a) America by Lewis Waterman
   b) France by Barthélemy Thimmonier
   c) China by Cai Lun

7   Rugby was invented
    a) In Scotland in 1457
    b) In America in 1891
    c) In England in 1823

8   In his model of a steam engine James Watt used
    a) A railway engine
    b) A washing-machine tub
    c) A giant syringe

9   The first telephone in 1861 included
    a) A dead man's ear
    b) A sausage skin
    c) A dead frog's skin

10 In 1895 Wilhelm Röntgen
    a) Was tried for witchcraft
    b) Invented rot-proof false teeth
    c) Discovered X-rays

11 Humans first began producing tools
    a) About 10.5 million years ago
    b) About 2.5 million years ago
    c) About 10,000 years BC

12 Who became angry about documents being typed?
    a) King Louis XV of France
    b) The Emperor of China
    c) Queen Victoria of England

Quiz Answers

PAGE 23

1 - b,   2 - c,   3 - b,   4 - b,   5 - b,   6 - b,
7 - b,   8 - c,   9 - b,   10 - c,
11 - a *and* c (trick question),   12 - b

PAGE 53

1 - c,   2 - a,   3 - c,   4 - c,
5 - a *and* b (trick question),   6 - a,   7 - c,   8 - b,
9 - b *and* c,   10 - c,   11 - a,
12 - a *and* b (trick question)

PAGE 91

1 - c,   2 - b,   3 - b,   4 - c,   5 - b,   6 - b,
7 - c,   8 - b,   9 - c,   10 - a, b *and* c (trick question),
11 - a,   12 - c

PAGE 124

1 - b,   2 - b,   3 - b,   4 - b,   5 - b,   6 - b,
7 - c,   8 - c,   9 - b,   10 - c,   11 - b, 12 - c

Index

If you have enjoyed this book, look out for:

THE SCIENCE MUSEUM BOOK OF AMAZING FACTS

SPACE

Anthony Wilson

For cosmic kids - a feast of weird
and wonderful facts about space!
The footprints that Neil Armstrong left on
the Moon will still be there in a million years.
Some of the atoms in your body are almost
as old as the universe.
When dinosaurs roamed the Earth, there were
only twenty-three and a half hours in a day.
Travelling at the speed of Concorde, a trip to the
nearest star would take 1.5 million years.

And look out for these AMAZING TITLES coming soon:

TRANSPORT
CONSTRUCTIONS
DISCOVERIES
EXPLORATIONS